Y0-BGD-284

# Experiments
# That
# Explore

## Oil Spills

# EXPERIMENTS
# THAT
# EXPLORE

# OIL
# SPILLS

**An Investigate! Book
by Martin J. Gutnik**

*Illustrated by Sharon Lane Holm
The Millbrook Press
Brookfield, Connecticut*

Cover photograph courtesy of Dale Guldan, The *Milwaukee Journal*

Photographs courtesy of Photo Researchers: pp. 2 (Pierre Burger), 9 (Peter B. Kaplan), 58 (Joe Munroe); National Center for Atmospheric Research/National Science Foundation: p. 24; Carolina Biological Supply: p. 30 (both); AP/Wide World: pp. 46, 48.


Cataloging-in-Publication Data

Gutnik, Martin J.
Experiments that explore oil spills / by Martin J. Gutnik.

p.   cm.—(An Investigate! Book)
Bibliography p.   Includes index.

Summary: The presence of spilled oil interrupts the natural cycle of life. We must act to prevent oil spills. Includes 11 experiments for young scientists.
ISBN 1-56294-013-9

1. Oil pollution of water.   2. Oil spills.   3. Marine pollution.
4. Water—Pollution.   I. Title. II. Series.
363.7     1991

# *Contents*

*Chapter One*
    Introduction 7

*Chapter Two*
    The Science of Water 13

*Chapter Three*
    The Ecology of Water 25

*Chapter Four*
    Water for Life 38

*Chapter Five*
    Oil Spills 47

*Chapter Six*
    The Future—What Can Be Done? 62

    *Conclusions to Projects* 64
    *Glossary* 65
    *For Further Reading* 69
    *Index* 70

*To Richard J. Cobb*

*Thanks to my wife, Natalie, for doing the primary research on this book.*

# 1

## *Introduction*

All of the water that the Earth will ever have is here now. There will never be any more or any less.

In all the Earth's waters, there is not one drop left that is pure. Even the snow at the top of Mount Everest has been found to have impurities.

The quality of the Earth's water is deteriorating. There are many major water pollution problems that demand immediate attention. One of these problems involves the transportation of petroleum products, specifically oil.

Modern society, in order to maintain its industrial plants and transportation systems, must use tremendous quantities of oil. In the moving of oil from its source to worldwide consumers, some of it is bound to spill into our waterways.

## THE HYDROLOGIC CYCLE

The *hydrologic cycle* (also called the *water cycle*) is the term used to refer to how water circulates around the globe. The Sun heats the atmosphere and causes surface water (water visible on the surface of the Earth, such as lakes and rivers) to *evaporate,* or turn into a gas. Water held within vegetation *transpires* (is released into the atmosphere). As the evaporated and transpired water rises, it cools and condenses to form clouds. These clouds are borne by the wind to all parts of the globe. When the clouds become heavy with moisture, they *precipitate* (release rain, snow, sleet, or hail). The precipitation replenishes surface water and groundwater (water beneath the surface).

## WATERY BIOMES AND ECOSYSTEMS

A *biome* is a community of living organisms within a single major ecological region, for example, a desert or a forest. An *ecosystem* is a smaller, more specialized community. It includes a variety of organisms that all interact to form a working unit. Examples of ecosystems are rivers, bogs, marshes, and ponds.

All of the planet's watery biomes and ecosystems are interrelated to each other and to the land-based biomes and ecosystems. Together they form the intricate web of living communities that make up this unique planet. And all of these communities are dependent on water in order to exist.

## PETROLEUM

Petroleum is a carbon-based, liquid fossil fuel found in the Earth. Millions of years ago, our planet was largely covered by shallow seas teeming with billions of microscopic plants and animals. When these organisms died, their bodies sank to the muck at the bottom of the seas. Because their bodies were buried in the muck, they did not totally *decompose* (break down). As the years went by, the seas became covered by layers of soil and rock. These layers compressed the bodies of the microscopic plants and animals until they formed the black liquid that today is called *petroleum,* or oil.

*Empty tankers leave New York City's harbor.*

Petroleum is a very useful product. It provides people with most of their fuel needs. It is used to make gasoline, diesel fuel, jet airplane fuel, and heating oil. It is also used in a multitude of other products, such as paint, that help make our lives better.

Geologists search for rock or soil formations that might contain oil. When they discover promising sites, wells are set up and drilling begins. If oil is found, it is pumped out of the ground and stored until it can be transported to a refinery. The refinery turns the raw oil into useful products. It is in the transportation of oil that major spillage problems occur.

Most oil is transported by pipeline or supertanker. When the tanker develops a leak, oil spills from the ship into the water. Oil is lighter (less dense) than water. Thus, when spilled, it floats on the surface of the water. Oil also spills from pipelines. Pipeline accidents are serious but, once found, are easier to clean up and contain than spills in the water. An oil spill in water has a much greater effect on the environment than one on land.

The presence of spilled oil interrupts the natural cycle of life in many ways. It prevents sunlight from penetrating the surface. This, in turn, prevents aquatic (water) plants from carrying on the important process of *photosynthesis,* or food making. It also blocks the natural exchange of gases, reducing the amount of oxygen (a gas that almost all living things need to exist) in the water. When animals such as birds, otters, and seals become covered with oil, they may suffocate or be unable to protect themselves from the elements. Birds covered with oil cannot fly, and when the oil is eaten, the tar balls poison the animal and it dies. The black substance also coats beaches and shorelines, making them ugly and destroying tide pools, where marine life thrives.

Clearly, we must act *now* to prevent oil spills from happening. Of course, accidents are bound to happen, and scientists are working to develop various ways of cleaning up spills that do occur. One cleanup method we'll talk about later uses microbes to "eat" the oil.

But what can we do about oil spills? The first step is understanding the problem. The projects in this book will demonstrate the makeup of water and the effect that oil spills have on water and the organisms that live in and depend on it. They will demonstrate just how destructive oil spills can be.

Some of the experiments in this book may be dangerous if special care is not taken by the experimenter. A few, particularly those requiring the use of heat or flame, should only be done with an adult present.

## THE SCIENTIFIC METHOD
## FOR SCIENCE PROJECTS

All science projects, which are investigations about nature and science, involve the scientific method of discovery. The scientific method is an orderly way of conducting research.

There are eight steps in the scientific method of discovery. Most scientists, as well as other researchers, follow these steps in one form or another. All of the investigations in this book will use the scientific method of research.

1. The first step of the scientific method is *observation*. The researcher uses his or her senses to find out all about an object or event. Not all the senses—sight, touch, hearing, smell, and taste—are used in every investigation. And, more often than not, special tools are used to enhance the senses. For example, microscopes enhance sight, and chemicals enhance or substitute for taste and sight. Researching a subject through the available scientific literature is also considered observation.

2. After an event or phenomenon has been observed, the investigator must classify what he or she has observed. *Classification* is putting together what you have observed into meaningful groups. For example, if you are observing amphibians, you would separate frogs, toads, and salamanders into three different groups. Later, you could further break these groups down into types of frogs, types of toads, and types of salamanders.

3 and 4. Once an event has been observed and classified, the researcher is now ready to make an *inference* or a *prediction*. An inference is an educated guess, based on what you have observed, about something that *has already happened*. Scientists quite often use inferences to study the causes of things: Why does cutting down rain forests cause environmental changes? How does acid rain affect lakes? Inferences are also useful in police investigations.

A prediction is an educated guess also, based on what you have observed, about something that *is going to happen*. Predictions are quite useful in the scientific investigation of space and weather patterns.

5. With the inference or prediction made, the researcher is now ready to formulate a *hypothesis*. A hypothesis is an inference or a prediction that can be tested. It is the focus of the scientific investigation. All things done while performing an experiment must relate back to the hypothesis. The hypothesis is usually in the form of an IF-THEN statement—for example, IF clouds are present, THEN rain is possible.

6. Once the hypothesis has been formed, it is ready to be *tested*. Testing the hypothesis is key to the scientific investigation. It is the experiment to be performed.

7. The testing completed, the *results* of the investigation must be recorded and analyzed.

8. Finally, with all of the above steps finished, the researcher is ready to draw a *conclusion*. The conclusion must always relate back to the hypothesis (for example: My hypothesis was correct or incorrect because . . .) and must state why your hypothesis was correct or incorrect. (Conclusions for most of the projects in this book can be found at the end of the book.)

Because quite often there are *variables* (outside conditions that may affect the results of the experiment), the conclusion may not always be valid. It is important to state the variables in your conclusion so that the results of your investigation will be easier to analyze.

Quite often, an investigation will produce results that do not completely answer the questions stated. This means it is time for further investigation. The researcher must restate his or her hypothesis and investigate again. Do not be disappointed if your results are unexpected. Some of the world's greatest scientific discoveries were the results of experiments that ''went wrong.''

# 2

## *The Science of Water*

Water ($H_2O$) is a *compound*. Each water molecule consists of two atoms of hydrogen and one atom of oxygen bound together chemically. Water is a colorless, odorless, clear liquid that covers 73 percent of the surface of the Earth.

Water heats up and cools down slowly, much more slowly than most other substances. This is because it is able to store large quantities of heat before its temperature begins to rise. This heat-storing characteristic is referred to as *specific heat*.

Due to water's slow fluctuation in temperature, a large body of water has tremendous influence on its surroundings. For example, the Great Lakes affect the weather conditions of the surrounding land areas.

Water is essential to life on Earth. Every living thing is dependent on water in order to exist. Many plants are 90 percent water (by weight). Many animals are 75 percent water. Jellyfish are 90 percent water. Most animals can survive much longer without food than water.

Water is a *universal solvent*. This means that most substances on Earth will dissolve (be physically broken up) in water. Mineral substances in the soil are dissolved in water and then absorbed by the roots of plants for use in photosynthesis.

Because water temperature changes much more slowly than the temperature of the air, the weather conditions in a watery environment are

## A Water Molecule ($H_2O$)

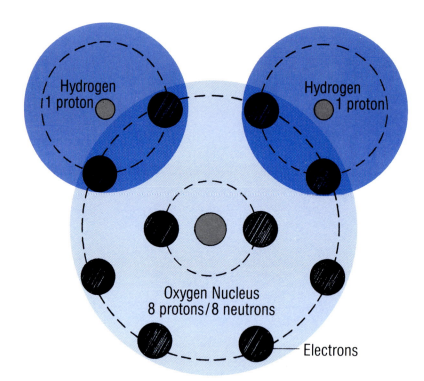

A water molecule consists of two atoms of
hydrogen and one atom of oxygen. They are
united in a chemical bond by the sharing of electrons.

much more stable and therefore more suitable to life than those on land. Plants and animals living in water are not subjected to rapid or extreme changes of temperature.

## THE THREE STATES OF WATER

Like all matter (matter is anything that has weight and takes up space), water can be found in three separate states, depending on its temperature. At temperatures between 32°F (Fahrenheit) and 212°F, water is a liquid. Below 32°F, water is a solid. And above 212°F, it is a gas.

When water freezes (becomes a solid), it expands and becomes lighter for each unit of volume. For this reason, a cubic foot of ice weighs less than a cubic foot of water. This is why ice rises to the surface of the water. When a body of water freezes, the ice on its surface protects the environment below from the cold temperatures. Thus, the life-forms beneath the ice do not freeze.

*Convection* is the main way heat moves in water. As the temperature of water falls, it becomes denser (heavier). In bodies of water such as lakes, ponds, oceans, and rivers, as the surface water becomes cooler, it gets denser. Because it is denser than the warmer water beneath, it sinks to the bottom and pushes the warmer (less dense) water up to the surface. This warmer water is then cooled, becoming denser and sinking to the bottom. In this manner, the entire body of water becomes gradually cooler, and extreme changes in temperature are avoided. This process is referred to as *fall-overturn*.

## Science Project #1— The Makeup of Water

### Materials Needed

a gas burner (use a stove at home or at school), a small pot, cold water

15

*Background for the Experiment.* The fuel from a gas burner is composed partly of hydrogen. When the flame is burning, the hydrogen from the flame will combine with oxygen in the air to form water that condenses on the cooler surface of the pot.

*Observations and Classifications.* The observations and classifications on water were obtained from reading about the properties of water.

*Inference.* There are several inferences that can be made from this reading. The important inference to us is that water is a compound made of two parts hydrogen and one part oxygen.

*Hypothesis.* If water is a compound, then it must be made up of at least two different kinds of atoms.

*Procedure.* Set a pot containing ice-cold water on the burner of a gas stove. Turn on the burner and observe the outside of the pot. Record your results.

*Results.* Droplets of water formed on the outside surface of the pot.

*Conclusion.* My hypothesis was correct because the hydrogen in the gas flame combined with the oxygen in the air to form droplets of water on the cooler outside of the pot.


## BREAKING DOWN WATER

Hydrogen and oxygen are gases that can be collected by bubbling them through water. Because the gases are lighter than water, they rise to the surface.

Oxygen gas supports burning (combustion). To test for oxygen, you will set a wood splint on fire, blow out the flames, and plunge the glowing splint into the test tube. If the splint re-ignites, it indicates the presence of oxygen. Hydrogen is more volatile than oxygen. When a glowing splint is brought near hydrogen gas, it will pop. An electric current passed through water (called *electrolysis*) will split the water molecules into separate atoms of hydrogen and oxygen.

## Science Project #2—
## The Electrolysis of Water

### Materials Needed

| | |
|---|---|
| 3 dry cell batteries | 2 test tubes |
| 2 iron nails | 2 wood splints and matches |
| electrical wire (bell wire) | a 1,000-ml or larger beaker of water |

*Observations and Classifications.*   Water is a compound made of two parts hydrogen and one part oxygen. All compounds can be broken down into their separate elements.

*Inference.*   Water should be able to be broken down into its two elements (atoms) of hydrogen and oxygen.

*Hypothesis.*   If an electric current is passed through water, then it will break the compound into its separate atoms of hydrogen and oxygen.

*Procedure.*   Hook up three dry cell batteries in a series current. (ASK YOUR SCIENCE TEACHER OR OTHER ADULT TO DO THIS FOR YOU.)

Attach an iron nail to each wire coming off the two end batteries. Do not hook up the series circuit until you are ready to conduct the experiment. Leave one of the connecting wires unattached.

Fill the 1,000-ml beaker and the two test tubes with water. Hold your thumb over the mouth of each test tube (one at a time), and place the test tubes upside down in the beaker of water. Remove your thumb

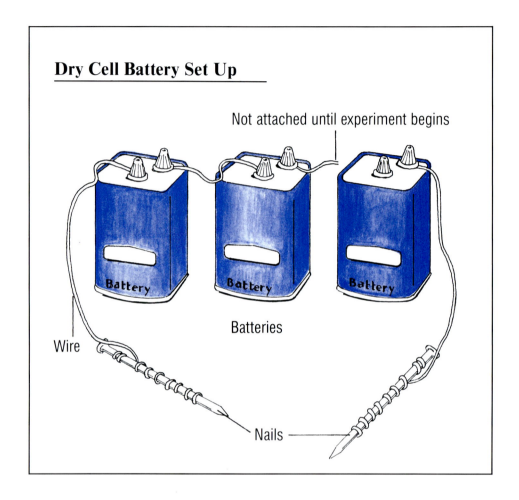

**Dry Cell Battery Set Up**

Not attached until experiment begins

Batteries

Wire

Nails

after the mouth of the test tube is under the water. By doing this, the water will stay in the test tube. Now put a nail, attached to each wire, under the test tubes and connect the unattached wire. Observe the test tubes and record the results.

Each test tube should fill with gas. The test tube collecting the hydrogen should fill twice as fast because there are two parts hydrogen to one part oxygen in each molecule of water.

# The Electrolysis of Water

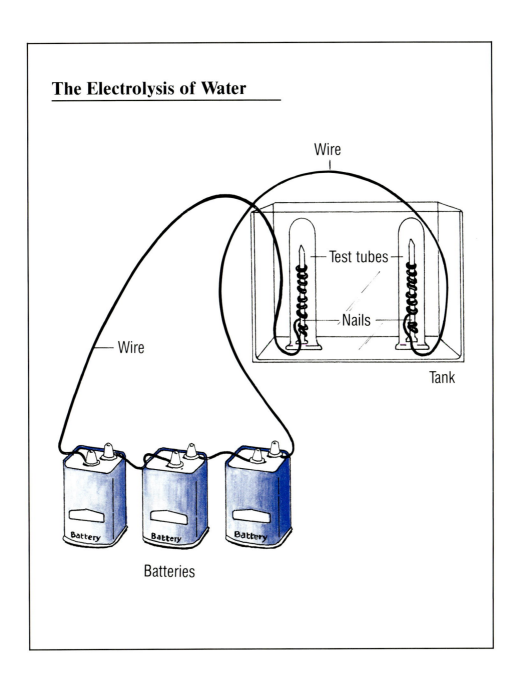

As the test tubes fill, the gases will push the water out of the tubes. After all the water is out of each tube, disconnect the wire between the batteries. Remove one of the test tubes in the same manner as you put it into the beaker of water (thumb over its mouth to prevent the gas from escaping).

Once the test tube is out of the water, have an adult light a match and start the wood splint burning. Now blow out the flame and plunge the glowing splint into the test tube. Repeat these steps with the other test tube. Record the results. Do this several times in each test tube.

***Results.*** 1. Each test tube filled with gas, which replaced the water in the test tube. 2. The splint re-ignited in one test tube. 3. The splint in the other caused a "pop."

See the CONCLUSIONS TO PROJECTS section from here on to find out what your conclusions should be.

## Science Project #3— Specific Heat

### Materials Needed

3 hot plates, alcohol burners, or stove burners

2 (500 ml) Pyrex or other heat-resistant beakers

3 thermometers

a 500-ml Pyrex flask

a 2-holed stopper to fit the 500-ml flask

a pair of tongs or hot pad

soil (250 ml)

water (250 ml)

***Observations and Classifications.*** As stated earlier, water heats up and cools down more slowly than most other substances. This is due to water's heat-storing abilities (specific heat).

***Inference.*** Water heats up and cools down more slowly than air or soil.

*Hypothesis.* If water, soil, and air are heated, then the water will heat up and cool down more slowly than the other two substances.

*Procedure.* Fill one of the beakers with 250 ml of water and the other with 250 ml of soil. Cork the flask with the two-holed stopper. Label the beaker with water *A*, the beaker with soil *B*, and the flask *C*. Put a thermometer in each beaker and the flask. Wait for 2 minutes and record the temperature of each substance on the charts in your results (see diagrams).

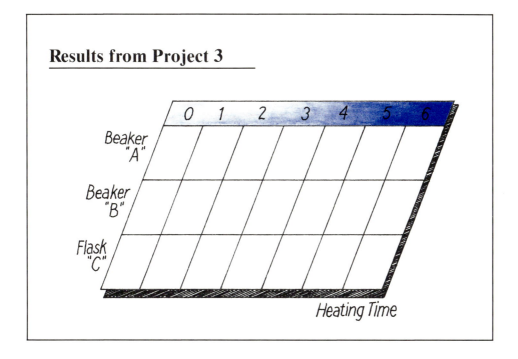

Place each beaker and the flask on a separate burner. Turn the burners on to the same temperature. Record the temperatures each minute as they

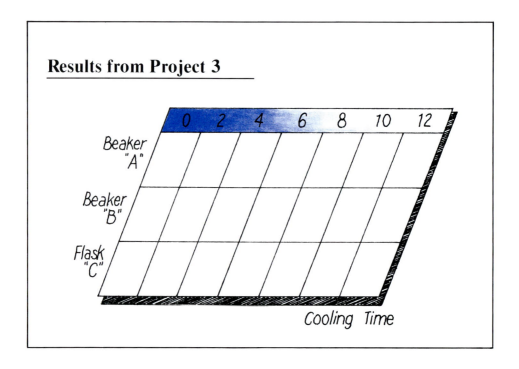

**Results from Project 3**

Beaker "A"

Beaker "B"

Flask "C"

0    2    4    6    8    10    12

Cooling Time

rise for the next 5 minutes. After 5 minutes, turn the burners off, and record the temperatures for the next 10 minutes as the substances cool.

## Science Project #4—
## Water as a Universal Solvent

### Materials Needed

2 (250 ml) beakers          ¼ cup salt
¼ cup sugar                 mixing spoon
balance scale

***Observations and Classifications.*** Almost all substances dissolve in water. Water dissolves minerals in soil into a nutrient-rich substance available to plants. Water percolating through the soil dissolves solid rock to form caves and underground rivers and lakes. When objects dissolve in water, they form solutions. A *solution* is a mixture of two or more substances. Neither substance is chemically changed in the mixture. Because of its universal solvency, water is never really pure but always in solution with one substance or another.

*Inference.* Since water is a universal solvent, it will form solutions with most other substances.

*Hypothesis.* If sugar and salt are mixed with water, then they will form solutions.

*Procedure.* Label one of your beakers *A* and the other *B*. Fill each beaker with 50 ml of water. Weigh the beakers on a balance scale. Add ¼ cup of sugar to beaker A. Add ¼ cup of salt to beaker B. Stir each mixture thoroughly, until the sugar and salt seem to disappear entirely. Place each beaker on a balance scale again to determine its weight.

Now place each beaker in a sunny place (for example, a windowsill) and let stand until all the water is evaporated. What happened? Record this in your results.

*Results.* 1. The sugar and salt dissolved in the water. 2. The beakers weighed more after the sugar and salt were dissolved. 3. After the water evaporated, the sugar remained at the bottom of beaker A. 4. After the water evaporated, the salt remained at the bottom of beaker B.

*Water is drawn up into clouds,*
*and later falls from them, as part*
*of the Earth's hydrologic cycle.*

# 3

## *The Ecology of Water*

Ecology is the study of how all living (biological) things interrelate to each other and their nonliving environment. All life on Earth exists within the *biosphere*. The biosphere is the portion of the Earth and its atmosphere that is capable of supporting life.

In order for life to exist, there must be a mixture of the four nonliving aspects of the biosphere—air, water, soil, and light energy. These four aspects are the physical elements of the biosphere.

All things on Earth and within the biosphere are interrelated and interconnected. Air moves about the globe by wind and a process known as convection. Soil is moved by wind, water, and organisms. Water moves by means of the hydrologic cycle. And all are affected by light energy.

The watery aspect of the biosphere is referred to as the *hydrosphere*. The hydrosphere is all the water and water vapor on Earth. The word *hydrosphere* comes from the Greek words *hydro,* meaning ''water,'' and *sphaira,* meaning ''sphere.''

Water, as stated earlier, moves about the Earth by means of the hydrologic cycle. As it moves, it absorbs and circulates heat from the Sun (solar energy). Because water is a universal solvent, it also contains many different chemical elements and compounds.

Geological studies of fossils indicate that throughout most of geologic time, all life on Earth existed within the oceans. It is only recently (within the last 280 million years) that life moved onto the land.

25

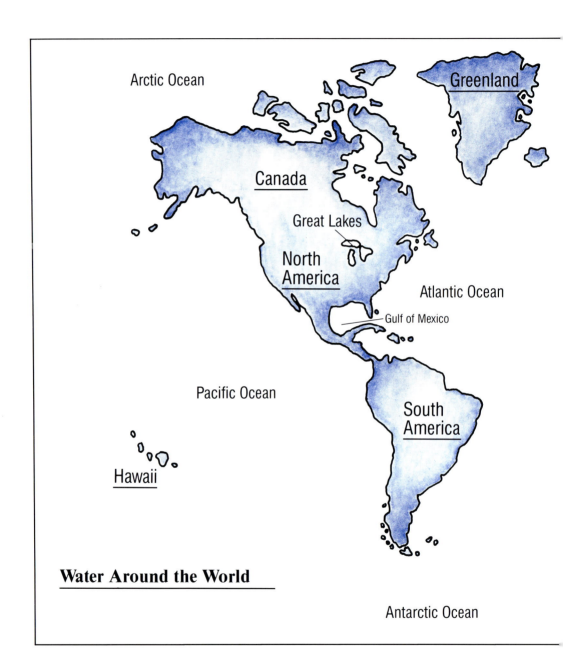

Arctic Ocean

Greenland

Canada

Great Lakes

North
America

Atlantic Ocean

Gulf of Mexico

Pacific Ocean

South
America

Hawaii

**Water Around the World**

Antarctic Ocean

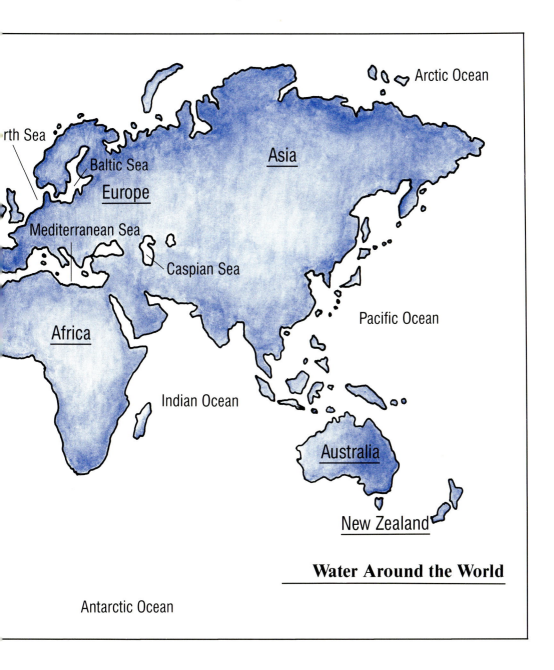

rth Sea

Arctic Ocean

Baltic Sea

Europe

Asia

Mediterranean Sea

Caspian Sea

Africa

Pacific Ocean

Indian Ocean

Australia

New Zealand

**Water Around the World**

Antarctic Ocean

## CLASSIFICATIONS
## OF WATER SYSTEMS

The Earth's water systems can be divided into two main groups, ocean waters and inland waters. Ocean waters cover approximately three quarters of the Earth's surface. They form an interconnecting hydrosphere that surrounds all the landmasses on Earth. Inland waters are found inside the borders of landmasses and usually flow toward the oceans. Inland waters usually contain fresh (sweet) water, while the oceans have salt water.

Ocean waters can be divided into separate ecological biomes according to depth, salinity (levels of salt in the water), and temperature. Individual ecosystems can be found in *estuaries* (places where rivers empty into oceans).

Inland waters can be divided into two subgroups, flowing water and standing water. Flowing water includes creeks, streams, brooks, and rivers. In all these waters, a moving current is usually apparent. Standing water usually consists of ponds and lakes.

All bodies of water—whether ocean or inland, standing or flowing— are part of biomes or ecosystems and support a complex ecological community. The entire community is interrelated by the *food web* in these systems. The food web is the relationship of organisms within a community and their dependence on one another for energy in the form of food.

## THE FOOD WEB IN
## AQUATIC SYSTEMS

In order for life to exist, there must be *interface,* a mixture of air, water, soil, and light energy. When all these aspects are present, an aquatic food web functions as follows: Light energy from the Sun is absorbed by green plants. Green plants are called the *producers* because they are the only organisms on Earth capable of producing food from solar energy. All other organisms (the *consumers*) are dependent on green plants, either directly or indirectly, for their food. In all water systems, the most important producers are *phytoplankton* (usually referred to as just *plankton*), microscopic plants that float and drift in the water. There are many other green plants in water systems, but plankton produce more energy than all the others combined.

The energy from green plants is transferred to all consumers in the food web by animals that eat only plants. These animals are called *herbivores*. The most important herbivores in water systems are *zooplankton* (microscopic animals). Other herbivores in water systems include small fish, clams, and so on.

Most of the larger fish and other aquatic organisms are *carnivores* (animals that eat only meat). There are many levels of carnivores in an aquatic ecosystem. Some carnivores only eat herbivores, others eat other carnivores, and still others eat both herbivores and carnivores.

Some aquatic organisms are *omnivores* (animals that eat both meat and plants). A good example of an aquatic omnivore is a turtle.

In an aquatic system, dead organisms and waste materials produced by organisms living in the water (together called *detritus*) sink to the bottom. At the bottom, they are decomposed by organisms such as bacteria, tubifex worms, and fungi. The decomposers break down the dead and waste material and return them to the ecosystem to be used as nutrients by plants for the manufacture of more food.


## Science Project #5—
## Creating an Aquatic Ecosystem

### Materials Needed

large fish tank (minimum 55-gallon capacity)

air pump to aerate the water

fine steel wool (obtainable from a hardware store)

light to go over tank

piece of glass measured and cut to the width of the tank and one-half the height

silicone sealant

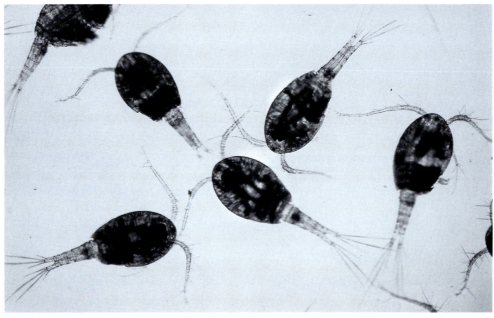

*Some of the most abundant creatures on Earth are the microscopic animals and plants that float in the world's oceans.*

flaked fish food
gravel
soil (for marsh or wetland area)
plant life for pond and marsh
    (elodea, myriophyllum, pond-
    weed or duckweed, cattails,
    marsh marigold, arrowhead,
    etc. Many of these plants may
    be obtained at local aquarium
    stores and garden centers.)
screen to cover tank

caulking gun
rocks
freshwater panfish, tadpoles, frogs,
    crayfish, salamanders, worms,
    turtle, minnows. (The ani-
    mals can be obtained from pet
    stores or from scientific supply
    houses such as NASCO or
    Fisher Scientific. Ask your
    science teacher for help in lo-
    cating these.)

*Observations and Classifications.*    Can be reviewed by reading the section on classification of water systems, p. 28.

*Inference.*    A fish tank can be set up to simulate a freshwater lake.

*Hypothesis.*    If we can set up an aquatic environment to simulate a freshwater system, then we can observe the food web in action.

*Procedure.*    Clean the fish tank thoroughly, using only fresh water and fine steel wool (no soap). It is important to remove all dust and other particles from the tank. This done, you are ready to create the pond and marsh portions of your aquatic model.

In order to separate the pond and marsh portions, the glass piece should be inserted and sealed into the middle of the tank, dividing the tank in half. Secure the glass in place by running a bead of silicone sealant along the bottom and sides of the tank. Carefully place the glass into the silicone bead, and, with your finger, smooth the silicone between the glass piece and the sides and bottom of the tank. The silicone, which takes 24 hours to cure, will waterproof these sections of the tank and prevent water from seeping through from one section to another.

Once the silicone has cured, cover the bottom areas of both sections with gravel. The gravel should be at least 3″ thick in the pond portion of the tank and 5″ thick in the marsh section. In the marsh area, the gravel should be sloped, with 3″ of gravel at the glass piece and 5″ at the back of the tank.

In the marsh section of the tank, add soil and rocks over the gravel. A special black, marsh-type soil can be obtained from some garden centers. Cover the gravel with the soil from 2″ below the glass divider, in a slight slope, to the rear of the tank. Place rocks in the pond and marsh for effect and to prevent soil runoff.

Before adding water to your tank, plant your vegetation. Wetland plants, such as cattails, marsh marigolds, and arrowheads, should be planted in the marsh area. Submerged vegetation, such as elodea, myriophyllum, water lilies, and duckweed, should be planted in the pond. (Water lilies should only be planted in very large tanks.)

Run a tube from your air pump into the tank. Run it under the gravel, leaving only the open tip to stick out of the gravel.

Fill the tank almost to the top with water. The water should be much deeper in the pond section. In the marsh area, some of the soil will be above the water level.

After allowing the water to stand in the tank for three days, you may add your animal life to the pond and marsh. Put in fish, tadpoles, frogs, and, if you wish, a turtle, crayfish, minnows, and other small aquatic animals. Quickly cover the tank with a framed screen top so that none of the animals will escape. Plug in your air pump to supply oxygen to the water.

Let the tank sit for two weeks, to allow time for the community to establish itself and for phytoplankton and zooplankton to develop.

After the two weeks have passed, observe the tank on a daily basis and record the behavior of the animals and plants in the food web. Try to draw a diagram, as shown here, of the food web in your aquatic environment. Do not dismantle your aquatic environment. It will be used later in this book for further experiments.

# Fish Tank with Rocks, Soil, Gravel, Plants, and Water

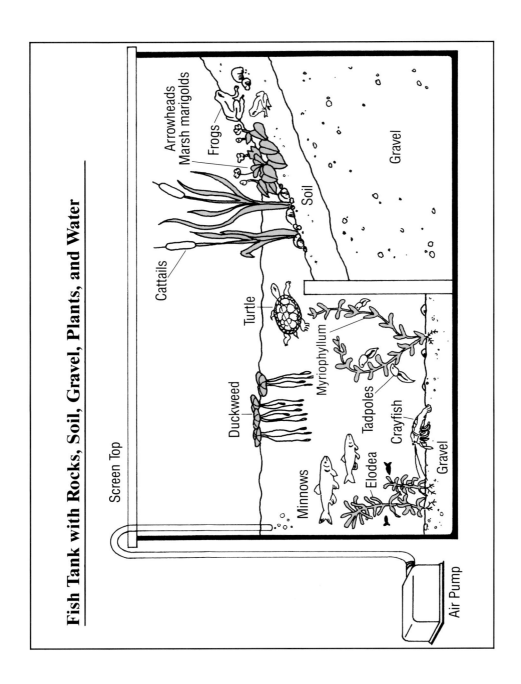

Screen Top

Arrowheads
Marsh marigolds
Frogs

Cattails

Soil

Gravel

Turtle

Duckweed

Myriophyllum

Tadpoles

Crayfish

Minnows

Elodea

Gravel

Air Pump

# Food Web in Aquatic System

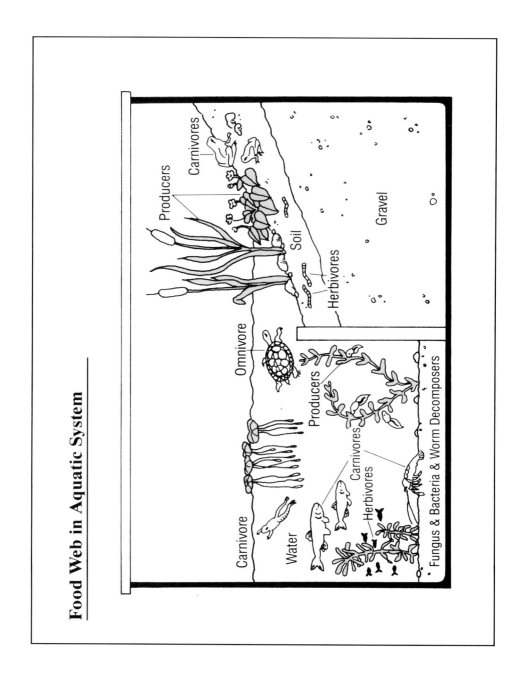

# Science Project #6—
# Plankton in a Pond Community

## Materials Needed

2 wide-mouth jars with covers
plankton net (to be made as part of the experiment)
medicine droppers
glass slides and cover slips for slides
compound microscope
manuals for identifying various kinds of phytoplankton and zooplankton
nylon stocking

wire clothes hanger
½-ounce lead fishing weights
rubber bands
needle and thread
small glass or plastic medicine bottle
string
wire cutters
metal ring (such as a washer or key ring)
epoxy glue

*Observations and Classifications.* Can be reviewed by reading this chapter on the ecology of water.

*Inference.* All ponds, both natural and people-made, contain plankton.

*Hypothesis.* If water samples are collected from a pond, then plankton will be in the samples.

*Procedure.* After your pond has been established for one month, a substantial plankton population should be flourishing. The purpose of this experiment will be to collect plankton from your pond and from a pond in your community and compare them.

The first thing you must do is build a plankton net. Bend a wire clothes hanger into a circle and close the circle by twisting the end. Use a wire cutter to cut off the excess hanger. Take an old nylon stocking and attach it to the circle of wire by sticking it all around the metal. (You may want an adult to help you with this.)

# Plankton Net and Collecting Bottle

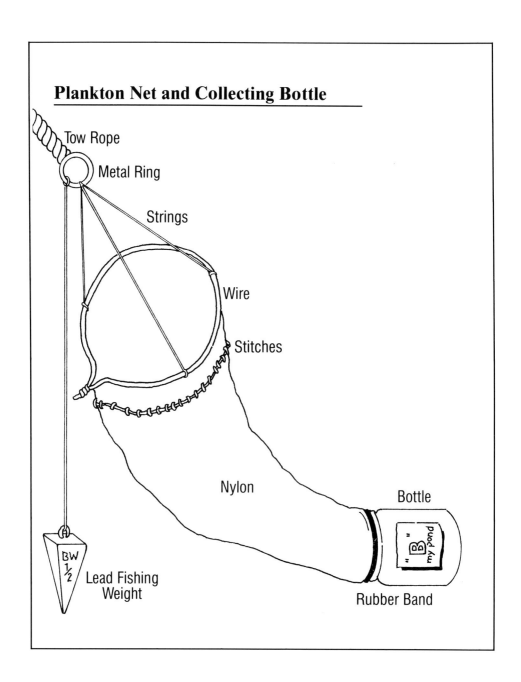

Tow Rope

Metal Ring

Strings

Wire

Stitches

Nylon

Bottle

BW 1/2

Lead Fishing Weight

Rubber Band

Cut off the end of the stocking at the toe, and with a rubber band attach the nylon to a small glass collecting bottle. Tie three short strings to the wire hoop. Use epoxy glue to hold them in place. Allow 24 hours for the glue to set. Attach the three strings to a small metal ring. Attach a ½-ounce lead fishing weight to the metal ring, and also attach a long section of string to the same metal ring. The long section of string will serve as your tow rope.

Label the two jars with covers. One jar should be labeled *A—Community Pond*, and the other *B—My Pond*.

Take jar A and your net to a pond in your community. At the shore, fill the jar half full of pond water. Cast the net out into the water and pull it through the water. Once you have retrieved the net and collecting bottle, take the bottle off the net and pour its contents into jar A. Cover the jar and return home. Put the jar in the refrigerator until you are ready to study its contents.

Take jar B and your collecting net to the pond you made in Science Project #5. Fill jar B half full of water. Dip the net into the pond area of the tank and pull it through the water. Remove the net and jar and empty the contents of the collecting jar into jar B. Cover jar B and put it in the refrigerator until you are ready to study its contents. Be sure to rinse the net well.

Take the jars out of the refrigerator and bring them to the table, where your microscope and manuals are set up. Uncover jar A and, with a medicine dropper, remove a small amount of water. Squeeze a drop of this water onto a glass slide and drop a cover slip over the water. DO NOT PRESS DOWN ON THE COVER SLIP. Place the slide on the stage of your microscope, focus, and study the organisms. Use your manuals to help identify the organisms you observe. Phytoplankton will be green because they contain chlorophyll. Record the names of the organisms and diagram them in your results. (Have a teacher or another adult help you learn how to use the microscope.)

Follow the same procedure as just stated above for jar B. Once done, compare and contrast the organisms found in the two different ponds.

# 4

*Water
for Life*

Water is essential to life on Earth. Plants use water to absorb dissolved minerals through their roots and carry on the process of photosynthesis. As part of the process, these plants transpire water. In other words, they release water through their *stomata* (pores on the underside of their leaves) into the air.

Transpiration, besides being part of photosynthesis, is also an essential part of the hydrologic cycle. Without transpiration, the world's rain forests would lose much of their water supply. It is transpiration that regulates the amount of rain that a rain forest receives. If transpiration did not take place, there would not be enough moisture in the air for rain to fall on the rain forest.

Animals use water to digest food and eliminate waste materials. Thus, water is essential to animals also.

It makes sense, then, that the availability of water greatly determines where life exists. The amount of water in any given area also determines what kinds of plants and animals will live in that area.

*Transpiration from leaves helps put moisture back into the air, where it can later rain down to nourish the rain forests.*

# Science Project #7—
# All Living Things Contain Water

*Materials Needed*

> potato
> potato grater
> 2-oz piece of steak
> celery stalk
> carrot slices
> oven
> gram scale
> dish

*Observations and Classifications.* Can be reviewed by reading the above text.

*Inference.* All living things contain water.

*Hypothesis.* If food products are collected and dried, then they will weigh less because of the loss of water content.

*Procedure.* Place a celery stalk and carrot slices on a gram scale and weigh them.

Next place them on a dish and let them stand for one week. Observe and record the results. Weigh the celery stalk and carrots. Record the weight.

Place the piece of steak on the gram scale and weigh it. Now put it in the oven, at 300°F, and heat it for 30 minutes. Take the steak out of the oven and weigh it. Record the results.

Grate a potato and observe the water content loss on the table. Pick up the shredded potato and squeeze it in your hand. What happens? Record your answer.

# Results of Project 7

| | Weight before Experiment | Weight after Experiment | Description |
|---|---|---|---|
| ● Celery | | | |
| ● Carrot | | | |
| ● Steak | | | |
| ● Potato | | | |

# PLANTS' DIFFERING NEEDS
# FOR WATER

As stated previously, all plants and animals are limited in terms of where they can live. Water must be available to them.

Through years of evolution, however, plants and animals have developed differing needs for water. Plants that live in dry ecosystems are called *xerophytes*. They have leaves, stems, and trunks that are designed to retain water. Many of these plants have root systems that can extend the search for water over large areas. Plants that live on land and receive an average amount of water are called *mesophytes*. These plants like water, but they also need to dry out between rainfalls. Plants that require water all the time are called *hydrophytes*. The roots or bodies of these plants must be constantly submerged in water.

A moisture-gradient box is designed to demonstrate how various plants are adapted to specific water conditions.

## Science Project #8—
## A Moisture-Gradient Box

### Materials Needed

large wooden waterproofed box (see Procedure), $3' \times 5' \times 1'$
waterproof material
gravel
water
pitcher
soil

plants: 4 each of xerophytes (cacti or succulents), mesophytes (marigolds or violets), and hydrophytes (wetland or submerged plants such as elodea or myriophyllum)
5 pieces of wood $2'' \times 4'' \times 3'$ long

***Observations and Classifications.*** See text above on plants' differing needs for water.

*Inference.*   Different types of plants require different water conditions.

*Hypothesis.*   If different plants need different amounts of water, then in a moisture-gradient box, wetland plants will survive at the bottom, woodland plants will survive in the center, and succulent plants will survive at the top.

*Procedure.*   Waterproof a large wooden box by lining it with plastic or some other waterproof material. Set the box up at an angle by using a two by four to support the back end of the box.

Once the box is set at a slant, lay a thin layer of gravel, about 1″ thick, over the bottom of the box. Cover gravel with 3″ of soil.

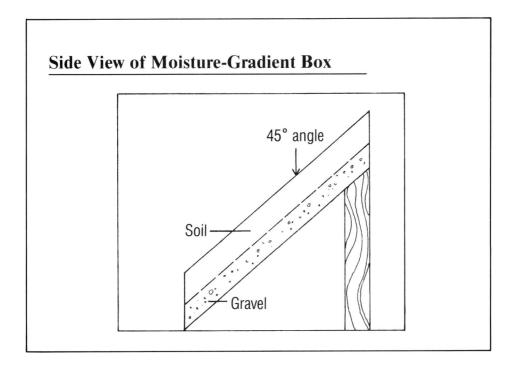

**Side View of Moisture-Gradient Box**

45° angle

Soil

Gravel

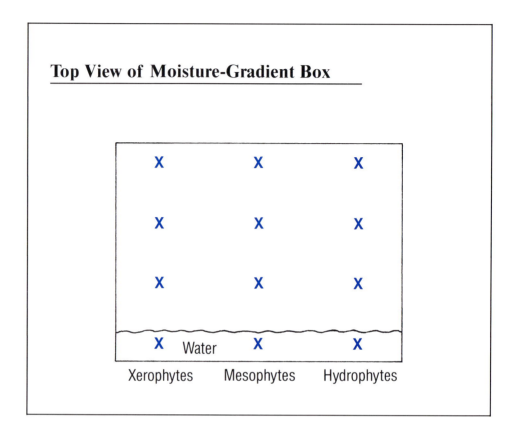

**Top View of Moisture-Gradient Box**

Xerophytes    Mesophytes    Hydrophytes

This done, you can now plant the vegetation in the box. On the left side of the box, plant a row (from the bottom of the box to the top) of xerophytes. In the middle, plant a row of mesophytes, and on the right plant a row of hydrophytes.

Pour water into the bottom of the box until a puddle of standing water remains. Repeat this until no more water is absorbed by the soil and there is some standing water left on the bottom of the box. Over the length of the experiment, *do not water the plants;* add water only to the bottom of the box.

Observe the box over a period of two weeks and record the results.

# Results of Moisture-Gradient Box Experiment

|  | In Water or Moist Soil | In Middle | At Top |
|---|---|---|---|
| Xerophytes | died | died | healthy |
| Mesophytes | died | healthy | dried up |
| Hydrophytes | healthy | dried up | dried up |

*Crude oil swirls on the surface of the waters of Prince William Sound after the disastrous* Exxon Valdez *spill.*

# 5

## *Oil Spills*

On March 24, 1989, the gigantic supertanker *Exxon Valdez* rammed into Blight Reef, releasing 11 million gallons of crude (unrefined) oil into Alaska's Prince William Sound. As the oil leaked out of the ship, it was carried on ocean currents and spread throughout the sound, then down the Gulf of Alaska to Cook Inlet and Kodiak Island. Four national wild-life refuges, three national parks, and the Chugach National Forest were affected. More than 1,200 miles of shoreline were blackened by oil.

The oil coated the water with a brownish-black film. Because it is lighter than water, the oil did not sink. The thin film of oil in Prince William Sound is estimated to have killed 90,000 to 270,000 birds, hundreds of sea otters, hundreds of thousands of shellfish, and many other plants and animals that live or depend on the sea.

On June 8, 1990, an explosion rocked the Norwegian supertanker *Mega Borg,* releasing more than 4.5 million gallons of light Angola crude oil into the Gulf of Mexico. As soon as the oil hit the water, it was set ablaze by the fire aboard the ship. The fire burned for days.

Although environmentally devastating to the surrounding area and the air, the fire was a mixed blessing. Had the oil not burned, it would have ravaged the estuaries on the Texas coast only 60 miles away. These estuaries are particularly rich in marine life.

*A red-necked grebe is covered with oil from the spill.*

## EFFECTS ON WILDLIFE

Both the *Exxon Valdez* and *Mega Borg* accidents had devastating effects on the wildlife in the areas they affected. In Alaska, bald eagles were poisoned as they hunted the oil-slicked waters for food. Birds such as murres, who preened themselves to restore their feathers, swallowed oil and died. As the surviving birds' feathers became soaked with oil, they could no longer act as insulation. This caused many birds to die of exposure to the elements. The toxic fumes from the oil caused many marine mammals, such as sea otters and seals, to suffer nosebleeds, emphysema (a lung disease), and blindness. Mammals that ingested oil suffered damage to their internal organs and eventually died.

Because of the *Mega Borg* oil spill in the Gulf of Mexico, the already threatened loggerhead sea turtle is now in serious trouble. Oil floating on the water has also destroyed millions of phytoplankton and zooplankton, thus disrupting the food web.

## THE CLEANUP

There were many delays after the *Exxon Valdez* spill. There was only one barge in the harbor capable of mopping up a spill and a serious lack of equipment. In addition, the size of the spill was huge. Finally, shortly after equipment and personnel were in place, bad weather broke up and spread the spill. Because of these reasons, the cleanup began too late, and the environmental effects were much more devastating than they should have been.

As soon as an oil spill occurs in American waters, the master of the ship is supposed to call the Coast Guard. The Coast Guard then sends a team out to investigate the spill and evaluate what equipment will be needed for containment and cleanup. The Coast Guard then notifies the spiller as to its responsibility. The spiller, by law, is responsible for cleanup of the spill.

In most areas where huge tankers load oil, cleanup equipment is *prestaged*. This means that equipment is already in position so that, if a spill occurs, it can be moved quickly to the disaster area. This was the case in the *Exxon Valdez* incident.

Once it has been noted that a spill has occurred, the cleanup crews go into action. Their first goal is containment of the spill. *Booms,* which are lengths of wood or other material with skirts that extend 6″ to 12″ above and below the water, are deployed around the spill. The skirts on the booms are designed to prevent the oil from spreading. If a spill occurs close to shore, booms are also deployed along the shoreline in order to protect these areas from the oil.

# Oil Skimmers

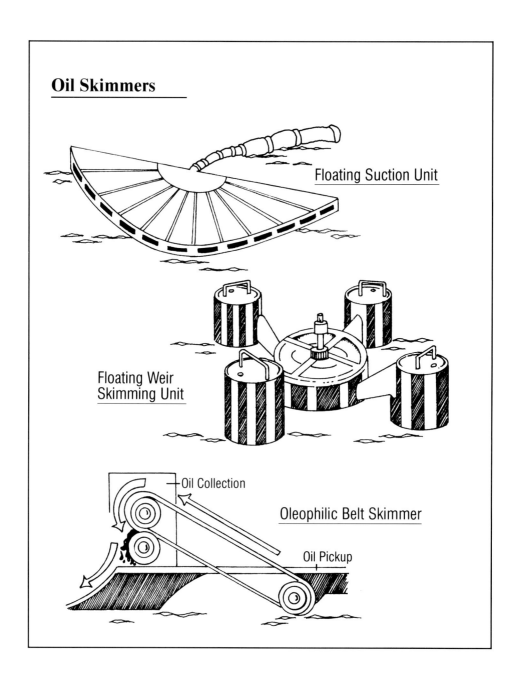

Floating Suction Unit

Floating Weir
Skimming Unit

Oil Collection

Oleophilic Belt Skimmer

Oil Pickup

With the booms in place, machines called *skimmers* are used to suck the oil off the water's surface. There are many different kinds of skimmers. Some skimmers have flat, absorbent surfaces on conveyor belts that collect the oil off the water's surface and pump it out to a ship or to a truck if the spill is close to shore.

If the oil spill is small, equipment called *pads* are placed on the spill. The pads act like gigantic sponges and soak the oil off the water's surface. Even with great efforts to contain a spill, however, once a massive spill occurs, there is not much that can be done to stop the environmental damage.

In the case of the *Mega Borg,* because of the fire, the oil slick was small compared with that of the *Exxon Valdez.* Skimmers were used on the mile-long oil slick. Also, for the first time on the open seas, oil-eating microbes were sprayed on the spill. These microbes are bacteria (one-celled organisms) that eat oil. They already exist in nature where oil naturally seeps into the ocean. Scientists have made a mixture of these bacteria, hydrogen, and phosphorus. The mixture is sprayed onto an oil spill, and the bacteria begin eating the oil. After the bacteria digest the oil, it is excreted back into the water as harmless fatty acids, which are eaten by small aquatic organisms. Oil-eating bacteria were also sprayed on Alaskan beaches after the *Exxon Valdez* spill reached the shore. They did help somewhat in the cleanup, but the spill was so vast that their effect was small.

Once an oil spill reaches the shore, as in the case of the *Exxon Valdez,* the cleanup becomes much more complicated. First the area where the oil has reached the shore must be isolated. Cleanup crews then come in with high-pressure hoses and attempt to wash the oil back to the water's surface, where it is soaked up by skimmers and pads. As the area is sprayed with the hoses, hot-water sprayers are used to clean the oil from the rocks.

The remaining oil, now a greasy slime, must be wiped by hand. The entire cleanup effort requires a lot of equipment and many people to assist and is only about 25 percent effective.

## Science Project #9—
## Oil and Aquatic Organisms

### Materials Needed

2 small fish bowls or large jars
4 elodea plants (can be taken from
    aquatic environment you built
    in Science Project #5 or
    bought at a local pet store)

motor oil (can be bought at a gas-
    oline station)
2 small lamps
gravel
black construction paper

***Observations and Classifications.*** As oil spreads over the water's surface, it prevents light and oxygen from getting to the aquatic organisms. Aquatic plants need light in order to carry on the process of photosynthesis, which produces food and oxygen. Aquatic animals need this food and oxygen in order to live.

***Inference.*** Aquatic plants and animals need light energy in order to live.

***Hypothesis.*** If an oil slick spreads across water, then it will prevent aquatic plants from carrying on the process of photosynthesis.

***Procedure.*** Put about 2″ of gravel in the bottom of each fish bowl. Label one fish bowl *A* and the other fish bowl *B*. Fill each fish bowl with water, and let it stand for one day. Plant two elodea plants into the gravel at the bottom of each bowl. Let the bowls stand for another day. On day three, cover the sides of each bowl with black construction paper. This will allow light to enter only from the top. Set a lamp over each fish bowl, and turn it on. The lamps will provide the light energy for photosynthesis. They should remain on for 13 hours a day. Again, allow the bowls to stand for one more day.

You will now simulate an oil spill over the water's surface. Open the can of motor oil and pour it into fish bowl B. Completely cover the surface of the water with oil. Let the fish bowls stand for one week. After that time, remove the black construction paper and observe the results by comparing fish bowl A and fish bowl B. (When photosynthesis occurs, oxygen bubbles appear in water.)

## Science Project #10— Containing an Oil Spill and Cleanup

### Materials Needed

aquatic ecosystem (from Science Project #5)

16 6″-long wood dowels, ⅛″ or ¼″ in diameter (caramel-apple sticks can also be used)

32 ¼″ × 6″ strips of tagboard or plastic (cardboard can be from shirts that have been sent to the laundry)

8 2″ × 2″ sponges (you may cut big sponges to this size)

Ping-Pong paddle

16 hooks and eyes (small enough to fit in ends of dowels)

a good water-resistant glue

⅛-ounce lead fishing weights

fish line

scrub brush

hot water

small screwdriver

large bowl with at least an 8″ diameter, filled with water

small fishnet

can of motor oil

rubber gloves (for wearing during the experiment to keep oil off hands)

plastic toy fish and frogs

*Observations and Classifications.* Can be reviewed by reading the section on oil spills in this book, pp. 47–51.

*Inference.* Oil spills are difficult to contain and to clean up.

*Hypothesis.* If oil is spilled into an aquatic environment, then it will be very difficult to remove and clean up.

*Procedure.* The first thing to do for this project is to build oil-spill containment booms. Each boom will be 6″ long and have two flaps. The

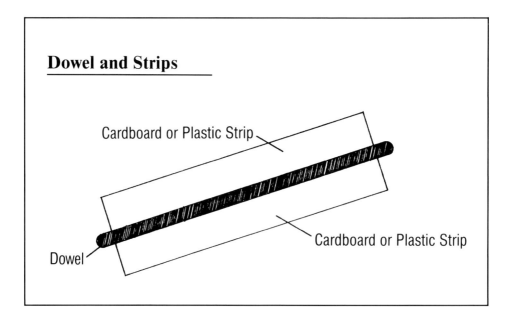

## Dowel and Strips

Cardboard or Plastic Strip

Cardboard or Plastic Strip

Dowel

booms will also have a hook on one end and an eye on the other end, so that they can be joined together.

Take one dowel and run a bead of glue in a straight line down one side of the dowel. Take one of your cardboard or plastic strips and set it into the bead of glue. Hold the strip in place until it stays. Allow this to set, then turn the dowel around and glue a second strip directly opposite it.

Once the strips have set, screw a hook and eye into each end of the dowel. This should be relatively easy, because the dowels are made of soft wood. Once you have done this, attach 6″ of fish line to each end of the dowel, and tie a ⅛-ounce fishing weight to each string. This should help keep the boom upright in the water and stop it from floating away.

After building your first boom, wait 24 hours for the glue to set, then test it to see if it works before building the other booms. Do this by laying the boom down in a bowl of water. The boom should float with one wing up and one beneath the water. If your boom does not lay this way, adjust your fish line and weight.

Once your boom works the way it is supposed to, build your other booms. When all the booms are built, be sure to test each one before you proceed with the experiment. Also, before spilling oil onto the water, remove fish, tadpoles, frogs, and any other animal life, and replace them with plastic animals.

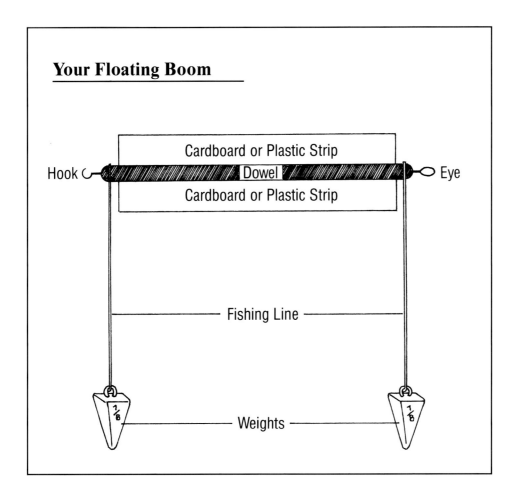

## Your Floating Boom

Cardboard or Plastic Strip

Hook   Dowel   Eye

Cardboard or Plastic Strip

Fishing Line

Weights

Imagine that a large supertanker is sailing across your aquatic ecosystem. It runs aground on a reef that tears a gaping hole in its side. Immediately, the oil starts to leak from the damaged ship.

Pour the oil onto the water of your aquatic ecosystem.

Your oil cleanup equipment is prestaged. The master of the ship has called for help. The equipment is rushed to the area, and the booms are set around the spill in order to contain the oil.

Set up your booms, hooking the ends together so that the entire oil spill is surrounded. What happens? Write down what you observe in your results.

Suddenly, a wind kicks up over the waters. The water starts to toss and turn, as the tempest worsens.

Use your Ping-Pong paddle to create waves and rough water in your aquatic ecosystem. Make the waves high enough so that the water splashes over the booms. What happens? Write down what you observe.

You cannot skim oil off the surface of the water during rough seas. It will have to wait until the weather improves. Your rough weather is going to last for two days.

Let the oil stay in your aquatic ecosystem for that long, every once in a while using your Ping-Pong paddle to make waves. Observe what happens during the two days.

After two days of rough weather, some of the oil should have washed up onto the shore. Did this happen in your experiment?

Now the cleanup begins. Use sponges to skim the oil off the surface of the water. Also, use your sponges to skim what excess oil you can off the beaches. Write down how the skimming went.

Check your aquatic system to see if any tar balls have formed. If so, clean them from the system by using a net to scoop them out.

Observe the water carefully to see if there is a brown, jellylike skin floating on its surface. This skin is called *mousse*. It is a mixture of oil and water, and it usually collects where the water is sheltered, for example, under rocks. If you find any mousse, remove it with a sponge.

On shore, check to see if the rocks are covered with oil. If so, remove them and scrub them off with hot water and a brush.

The cleanup is now finished. Check your aquatic system to observe any effects the oil spill had on the vegetation. What effects do you think it would have had on the live animals, had you not removed them? Scoop some water out with the plankton net from Science Project #6 and, using the methods in Science Project #7, see how the spill affected the plankton. Record your observations.

***Results.*** 1. The oil spill was contained. 2. The oil splashed over the booms and was spread everywhere by the rough water. 3. The skimming . . . (you finish this statement).

## DRILLING FOR OIL
## IN THE OCEANS

For years, we have heard environmentalists argue that offshore oil rigs are damaging to marine environments. As the noisy drill bit turns, it brings up wastewater filled with poisons that are eaten by the animals that live in the area. Environmentalists also fear that the muds discharged by the rigs will bury the worms and other small creatures that live on the ocean floor.

Some environmentalists now also feel that there is additional danger in the transportation of oil to shore. Pipelines can leak, tanks do leak, and the shoreline itself—which is home to many unique ecological communities—is being destroyed.

Oil companies argue that the oil rigs have proved useful to the marine environment. Abandoned rigs serve as excellent artificial reefs that attract fish and other aquatic organisms. There are approximately 4,200 oil and gas rigs in the world today—plenty of artificial reefs for the foreseeable future!

*An offshore oil rig.*

# Science Project #11—
## Making an Artificial Reef

*Materials Needed*

aquatic environment (from Science Project #5)

pieces of scrap wood with nails or metal attached so that they will sink to the bottom of the water

pieces of scrap iron (available from any junkyard)

10 minnows (any variety; may be obtained at a bait store)

10 crayfish (may be obtained from a scientific supply house; ask your teacher for help in locating these)

all other organisms from the aquatic environment (if necessary supplement with more aquatic plants, such as elodea)

*Observations and Classifications.* Any angler knows that the best place to catch many kinds of fish is around a structure. A structure in the water serves many purposes. It attracts aquatic plants such as algae, which grow on the sides of the structure. Small aquatic animals, such as zooplankton, worms, crustaceans, and small fish, are attracted to the structure because they eat these aquatic plants and each other. The smaller fish and crustaceans use the structure for shelter and for hiding from bigger fish, which are attracted to the reef because they eat the smaller fish, worms, and crustaceans.

*Inference.* Debris left in the water eventually will turn into a reeflike structure, which will attract aquatic life.

*Hypothesis.* If debris is placed on the bottom of a body of water, then it will attract aquatic plants and animals.

*Procedure.* Take pieces of scrap wood and metal, and drop them to the bottom of the pond area in your aquatic environment. Make the pile sub-

# Chart of Results

| Behavior Around Artificial Reef | |
|---|---|
| Minnows | |
| Crayfish | |
| Larger Fish | |
| Insects | |
| Worms | |
| Etc. | |

60

stantial, at least 18″ high and 24″ wide. Try to leave spaces in the debris so that small aquatic animals will be able to go in and out.

Add the minnows and crayfish to the pond.

Let the debris stand for two weeks. Then observe the debris pile for a week's time. Record the behavior of the organisms.

*Results.*   1. Aquatic plants have begun to grow on the debris. 2. Some of the floating elodea plants have gotten caught in the debris and are growing there. 3. Behavior around the artificial reef . . . (you fill in the rest of this statement).

# 6

## *The Future –*
## *What Can Be Done?*

There is one basic fact that all the experts and environmentalists can agree on. Once a huge oil spill occurs, there is very little anyone can do to stop the environmental devastation. It is apparent, therefore, that the answer to preventing environmental disasters such as occurred with the *Exxon Valdez* accident is to prevent the spill.

The U.S. Congress has attempted to do this by enacting strict legislation concerning oil company tanker operations. This legislation requires thorough inspection and control of tanker operations and crew training.

Congress is also moving toward the passage of an oil-spill law that will include a cap on how much a company can be forced to pay for cleanup of a spill and a requirement for double hulls for all new tankers.

For its part, the oil industry has proposed a multimillion dollar program to establish five coastal cutters to respond in the event of oil spills. The companies are still debating the double-hull requirement, many saying that the cost of building double-hulled tankers is too high.

Environmentalists heavily support the double-hull requirement, citing the *Nancy Gaucher* incident. The *Nancy Gaucher,* a supertanker, smashed onto the rocks in the Detroit River's Livingston Channel. Not one drop of its 4,275 tons of oil leaked out. The *Nancy Gaucher* has a double hull.

Many members of Congress, however, want to wait for the results of a study being done by the National Academy of Sciences. This study is examining whether double hulls can be economical.

The fact is, however, that the money Exxon spent for the cleanup of the *Valdez* spill would have paid for double hulling almost 100 tankers. Combine this with the money spent on the *Mega Borg* incident and others, and it seems clear that it would be more economical for the oil companies to double hull their fleet than to pay for the cleanup of yet more spills, which are bound to occur.

Another way of reducing oil damage to the environment is through conservation. If we reduce our need for oil, there will naturally be less drilling, shipping, and, as a result, fewer spills.

Whatever the results of the congressional debates, something must be done to prevent oil spills in the future. The Earth's environment is a fragile balance of interrelated systems. The more often these systems are disturbed, the more likely that permanent damage will be done.

**2.** My hypothesis was correct because the water was broken down into separate atoms of hydrogen and oxygen.

**3.** My hypothesis was correct because the water heated up and cooled down more slowly than the other two substances.

**4.** My hypothesis was correct because both the sugar and the salt formed a solution in the water.

**5.** My hypothesis was correct because the food web was observed and recorded.

**6.** My hypothesis was correct because there were both phytoplankton and zooplankton in the ponds.

**7.** My hypothesis was correct because the food products all contained water.

**8.** My hypothesis was correct because each plant adapted to its own specific water needs.

**9.** My hypothesis was correct because the aquatic plants could not carry on the process of photosynthesis.

**10.** My hypothesis was correct because oil spills are extremely difficult to remove and clean up.

**11.** My hypothesis was correct because the debris on the bottom of the aquatic environment did attract aquatic life.

# *Glossary*

**Aquatic.** Referring to water, for example, water plants and water animals.

**Atmosphere.** The primary layers of gases surrounding the Earth. The atmosphere extends approximately 600 miles above the Earth's surface.

**Biological.** Refers to any living thing.

**Biome.** The community of living organisms of a single major ecological region, such as a rain forest or a desert.

**Biosphere.** The portion of the Earth and its atmosphere that is capable of supporting life.

**Boom.** A mechanism designed to float on the water's surface and contain an oil slick.

**Carnivore.** Any animal that eats only meat.

**Classification.** The putting of objects into groups according to their properties.

**Compound.** Any new substance formed from two or more separate substances. For example, water is a compound of hydrogen and oxygen.

**Conclusion.** The gathering and interpreting of data in a scientific experiment, in order to find out if your hypothesis was correct or incorrect.

**Consumer.** Any organism that is not a producer of food but uses the food that the producers—green plants—make.

**Convection.** The movement of air or water created by temperature. Cold water sinks and pushes the warmer water to the surface.

**Density.** The mass of a substance, per unit of volume.

**Detritus.** The remains of creatures that have died plus waste materials produced by living creatures, which becomes food for certain species such as worms and flies.

**Ecology.** The study of how all living things interrelate with each other and their nonliving environment.

**Ecosystem.** A specialized community, including all the component organisms, that forms an interacting system, for example, a bog.

**Electrolysis.** The splitting of water molecules into their component parts by running an electric current through water.

**Element.** Any substance made up of one kind of atom, such as oxygen. Elements cannot be chemically broken down.

**Evaporation.** Surface water going into the air as a gas (water vapor).

**Fall-Overturn.** The process of convection in water. It causes cool water to sink and warmer water to rise to the surface. In this manner, the entire body of water becomes gradually cooler.

**Food Web (Chain).** The special relationships of the organisms within a community and their dependence on one another for energy in the form of food.

**Fossil Fuel.** Any fuel formed from the fossil remains of plants or animals, for example, coal, oil, and natural gas.

**Herbivore.** Any organism that eats only vegetation.

**Hydrogen.** A gas contained in the compound water. Each water molecule is two parts hydrogen to one part oxygen ($H_2O$).

**Hydrologic Cycle.** The regular movement of water from the atmosphere, by precipitation, to the Earth's surface and its return to the atmosphere by evaporation and transpiration.

**Hydrophyte.** A plant that grows wholly or partly immersed in water.

**Hypothesis.** An educated guess that is formed from the premise for an experiment or problem. An inference or prediction that can be tested.

**Inference.** An educated guess based on what has been observed.

**Interface.** In the biosphere, where air, water, soil, and light energy form a proper mix to allow for the existence of life.

**Mesophyte.** A plant that grows in environmental conditions that are average in terms of moisture.

**Microbe.** An organism so small that it cannot be seen without the aid of a microscope.

**Mousse.** A thick brownish mixture of oil and water that floats on the surface of the water.

**Nutrients.** All substances that an organism obtains from its environment (except for water and gases).

**Observation.** A method, using the five senses, of studying scientific or natural phenomena.

**Oil.** A common name for petroleum. See also **Petroleum.**

**Omnivore.** Any animal that eats both plants and meat.

**Organism.** Any living thing.

**Oxygen.** A gas that makes up about 21 percent of our air.

**Pad.** An absorbent material used to soak up an oil spill.

**Petroleum.** A carbon-based liquid fossil fuel found in the Earth.

**Photosynthesis.** The process by which green plants make food (the simple sugar glucose) and give off oxygen in the process.

**Phytoplankton.** Microscopic plants that float in the water.

**Precipitation.** Water falling from the atmosphere in such forms as rain, sleet, and hail.

**Prediction.** An educated guess, based on observation, about something that is going to happen.

**Prestaged.** Oil-spill cleanup and containment equipment that is in place in high-risk areas.

**Producers.** Organisms that make their own food, for example, green plants.

**Reef.** A strip or ridge of rocks, sand, coral, or other substance that rises to or near the surface of a body of water.

**Results.** The record of what happened during a scientific investigation or experiment.

**Salinity.** The level of salt in water.

**Series Circuit.** An electrical circuit connected so that current passes through each circuit element in turn, without branching.

**Skimmer.** A mechanism used in oil cleanup operations. It provides a means of removal and recovery without changing the physical and chemical properties of the oil.

**Solution.** A balanced mixture of two or more substances.

**Specific Heat.** The quality of water that allows it to store large quantities of heat before its temperature begins to rise.

**Stomata.** Openings, such as pores, on the underside of a plant's leaves that release oxygen and water during the process known as transpiration and take in carbon dioxide.

**Transpiration.** The process whereby oxygen in the form of a gas is released into the air through the leaves of plants. The oxygen is a by-product of photosynthesis, the food-making activity of plants.

**Universal Solvent.** Refers to water's ability to dissolve most substances on Earth.

**Variables.** Differing conditions that might affect the outcome of a scientific investigation or experiment.

**Water.** A colorless, odorless, tasteless liquid made up of the compound of two parts hydrogen and one part oxygen ($H_2O$).

**Xerophyte.** A plant that can grow under arid (dry) conditions.

**Zooplankton.** Microscopic animals that float in water.

# *For Further Reading*

Brown, Joseph E. *Oil Spills*. Putnam: New York, 1978.

Cross, Wilbur. *Petroleum*. Children's Press: Chicago, 1983.

Gutnik, Martin J. *How to Do a Science Project and Report*. Franklin Watts: New York, 1980.

Kraft, Betsy H. *Oil and Natural Gas*. Revised edition. Franklin Watts: New York, 1982.

Pampe, William R. *Petroleum: How It Is Found and Used*. Enslow: New Jersey, 1984.

Twist, Clint. *Rain to Dams: Projects with Water*. Franklin Watts: New York, 1990.

Woods, Geraldine and Harold Woods. *Pollution*. Franklin Watts: New York, 1985.

# *Index*

Air, 25, 28
Alaska, 46-49, 51
Algae, 59
Aquatic ecosystems, 29, 31-32
Artificial reef, 59, 61

Bacteria, 29, 51
Bald eagles, 48
Biome, 8, 28
Biosphere, 25
Birds, 10, 47, 48
Booms, 49, 51, 54-56

Carnivores, 29
Chugach National Forest, 47
Classification, 11 (*see also* Science projects)
Clouds, 8, 24
Coast Guard, 49
Compound, 13
Conclusion to projects, 12, 64
  (*see also* Science projects)
Congress of the United States, 62
Conservation, 63
Consumers, 29
Convection, 15, 25
Cook Inlet, 47
Crustaceans, 59

Decomposition, 8
Density
  of oil, 9
  of water, 15
Depth of water, 28
Detritus, 29
Double-hull requirement for tankers, 62-63

Ecology, defined, 25
Ecosystems, 8, 28-29, 31-32
Electrolysis of water, 17-20
Estuaries, 28, 47
Evaporation, 7
*Exxon Valdez*, 46-49, 51, 61, 62

Fall-overturn, 15
Fish, 29, 59
Flowing water, 28
Food web, 28-32
Fossil fuel, 8, 9
Fossils, 25
Fresh (sweet) water, 28
Fungi, 29

Geologists, 9
Great Lakes, 13
Green plants, 28-29

Groundwater, 8
Gulf of Alaska, 47
Gulf of Mexico, 47, 49, 51

Herbivores, 29
Hydrogen, 13, 16-18, 51
Hydrologic cycle, 7-8, 24, 25, 38
Hydrophytes, 42, 44
Hydrosphere, 25, 28
Hypothesis, 11 (*see also* Science projects)

Ice, 15
Inference, 11 (*see also* Science projects)
Inland waters, 28
Interface, 28

Kodiak Island, 47

Light energy, 25, 28
Livingstone Channel, 62
Loggerhead sea turtle, 49

Materials for science projects, 15, 17, 20, 22,
  29, 31, 35, 40, 42, 52, 53, 59
*Mega Borg,* 47-49, 51, 63
Mesophytes, 42, 44
Microbes, oil-eating, 10, 51
Microscope, 37
Moisture-gradient box, 42-44
Mount Everest, 7
Mousse, 56

*Nancy Gaucher,* 62
National Academy of Sciences, 63

Observation, 11 (*see also* Science projects)
Ocean waters, 28
Oil
  density of, 9
  ocean drilling for, 57-61
  spills (*see* Oil spills)
  transportation of, 7, 9

Oil-eating microbes, 10, 51
Oil spills
  cleanup of, 10, 49-51
  damage caused by, 9-10, 47-49
  prevention of, 62-63
  science projects on, 52-55
Omnivores, 29
Organisms, 25
Otters, 10
Oxygen, 13, 16-18

Pads, 51
Petroleum
  history of, 8
  uses of, 9
  (*see also* Oil)
Phosphorus, 51
Photosynthesis, 10, 13, 38, 52, 53
Phytoplankton, 29, 32, 35-37, 49
Pipeline accidents, 9, 57
Plants, 10, 28-29, 42-44
Pollution, 7
Precipitation, 8
Prediction, 11
  (*see also* Science projects)
Prestaged equipment, 49, 56
Prince William Sound, 46, 47
Procedures, 16-18, 21-23, 31-32, 35, 37, 40,
  43-44, 52-55, 59, 61
Producers, 28

Rain forests, 38, 39
Refineries, 9
Results of investigation, 12 (*see also* Science
  projects)

Salinity of water, 28
Salt water, 28
Science projects
  aquatic ecosystem, 29, 31-32
  artificial reef, 59, 61
  conclusions to, 12, 16, 64

Science projects (*continued*)
  electrolysis of water, 17-20
  living things contain water, 40
  makeup of water, 15-16
  moisture-gradient box, 42-44
  oil and aquatic organisms, 52
  oil spill containment and cleanup, 53-57
  plankton in pond community, 35, 37
  specific heat, 20-22
  water as universal solvent, 22-23
Scientific method of discovery, 10-12
Scientific supply houses, 31
Seals, 10, 48
Sea otters, 47, 48
Shellfish, 47
Skimmers, 51
Soil, 25, 28
Solar energy, 25, 28
Solutions, 23
Specific heat, 13, 20-22
Standing water, 28
Stomata, 38
Sun, 25, 28
Surface water, 7, 8

Tankers, 9, 47-49, 62-63
Temperature of water, 13, 15, 28
Testing the hypothesis, 12 (*see also* Science projects)

Transpiration, 7, 38, 39
Tubifex worms, 29
Turtles, 29, 49

Universal solvent, water as, 13, 22-23, 25

Variables, 12

Water
  breaking down, 16-23
  classification of system, 28
  density of, 15
  electrolysis of, 17-20
  makeup of, 15-16
  oil spills and (*see* Oil spills)
  plants' differing needs for, 42-44
  quality of, 7
  specific heat of, 13, 20-22
  temperature of, 13, 15, 28
  three states of, 15-16
  as universal solvent, 13, 22-23, 25
Water cycle, 7-8
Wildlife, effects of oil spills on, 9-10, 47-49
Wind, 8, 25
Worms, 59

Xerophytes, 42, 44

Zooplankton, 29, 32, 49, 59